Website Development Methodology

Putting it all Together
Team Building for Winners

By Robert Villegas

Website Development Methodology
By Robert Villegas

Published by Document Services International

www.documentservicesinternational.com

ISBN-13: 978-1517463717

ISBN-10: 1517463718

Table of Contents

I. Introduction

> *"In reengineering projects (and in most application development projects), methodologies are like Robert Louis Stevenson's 'bright face of danger.' They are attractive insofar as they provide consistency and direction, and they are useful when the environment is stable and there is already expertise in applying a methodology. However, the search for methodology has been the death knell for many a project. Too often the quest for defining how to do the work overtakes actually doing the work, and the project goes nowhere."* [1]

In today's corporation, especially those involved in the web design industry, everything is marketing. Effective strategic marketing requires the ability to differentiate the website development organization and its deliverables from those of the competition. All business processes should be designed to facilitate the provision of the highest quality services or project deliverables to the customer — and thereby create the marketing story that brings customers to the door. If the process does not deliver better products and services, speed-to-market and project completion under budget, then it must be reengineered until it does.

The goal of this document is to provide a framework for building winning web sites and for keeping the "how to" from overtaking the work. We seek to provide a basic methodology that will enable consistent practices and procedures for the development and integration of E-commerce Web Sites.

In the sports arena, we marvel at the site of the rare super athlete whose on-field exploits greatly eclipse the performances

---

[1] Successful Reengineering by DaNiel P. Petrozzo and John C. Stepper Page 190

of other players. Though these exceptional superstars will live in the history books, with rare exception, their teams are often relegated to the position of "also-rans" in their sport. This is because most of those teams were not able to develop and exploit the synergies that proceed from a superlative *team* effort. Superlative athletes like Mark McGuire, Charles Barkley and Dan Marino are good examples. Their teams serve as poignant reminders that, no matter how great the individual players, if they do not play well *together*, success is not possible.

In like fashion, whether we come to a web design project from the technical, design or business discipline, we tend to view our specialized roles in the team dynamic as more important than that of the other disciplines which we regard as supporting our efforts. Effective project management requires the active and equal participation of every individual and every design discipline. When each team member understands that our unique and superlative contribution is also part of a team effort, and that the team must succeed in order for the individual to succeed, you will develop prize-winning customer deliverables. And as with all professional teams, you must recognize that the individual's role is to contribute his very unique capabilities with a maximum of effort and diligence. When the team functions like a well-oiled machine, so to speak, then you will realize a winning outcome.

The entire range of your corporate professionals, whether they are consultants, technologists, graphic designers, project managers, executive management or administrative staff will find this document helpful in identifying key roles and responsibilities for the fulfillment of Client web design requirements.

As you know, Web Site development does not exist in a vacuum. Many companies over the last few decades have independently evolved Web Site development principles and methods "on the job," while they produced the web sites and E-commerce sites that are found today on the World Wide Web. These methodologies have gravitated toward basic processes like Requirements Gathering, Design, Development, Testing and Implementation/Distribution. We have analyzed many of these methodologies and have striven to provide a Web Site Development Methodology based upon tested design principles that will enable your company to provide the highest quality in state-of-the-art E-commerce Web Sites for your Clients and partners.

2. Development Environments

Development environments range across the full spectrum of work environments gaining the benefit of accommodating the needs of the best professionals in the technology fields.

a. Development Center

The development center environment is a full-fledged center whose only goal is the development of Web Site and Client systems. (See next page) It is designed to enable E-commerce professionals to have access to the latest in development tools and hardware while working in a corporate and team environment.

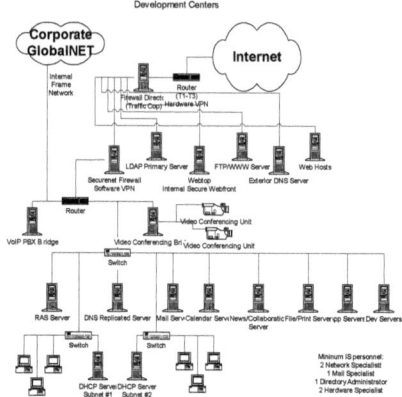

Figure 1. The Development Center Environment

b. The Remote Environment

The remote environment enables the use of regional or local talent in some of the best technology centers around the world. This environment enables technology workers to gravitate toward those centers where the technology jobs reside and it enables the design firm to effectively contract with them for their services.

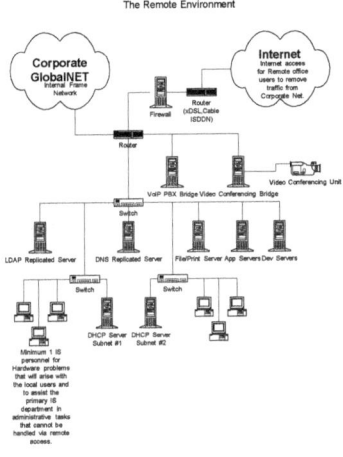

Figure 2. The Remote Environment

c. The Virtual Employee

The virtual office enables highly talented employees to work from home or a local office. This environment is useful for needed employees who would prefer not to relocate or those who may be travel-challenged.

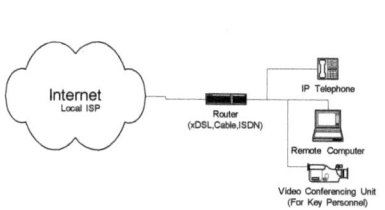

Figure 3. The Virtual Employee

3. Putting It All Together—Team Building for Winners

The Venn diagram below (Figure 4) illustrates the relationships among the three delivery models. The yellow regions represent the resources that are put into action during the design and planning of a project. The red regions illustrate the convergence of two areas into methodology and program management work products. The blue region illustrates the successful deliverable to the Client: a state-of-the-art E-commerce solution.

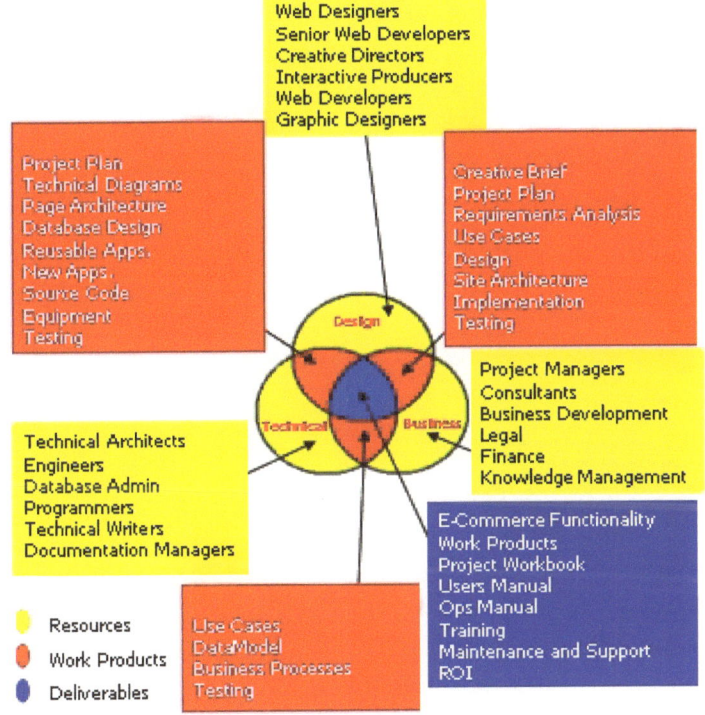

Figure 4. Venn Diagram illustrating deliverables.

As you implement this methodology in your organization, the three business units interact using sound business principles and planning methodologies to produce the final deliverable – a web site for your client.

a. Business

The Business Unit consists of the Business Development departments and all support functions. The goal of the Business Unit is to maintain and advance the Client relationship as well as provide the resources needed by the entire organization. These include Business Development and related activities, Knowledge Management, Human Resources, Program and Project Managers and Legal and Finance.

The Business Unit Interfaces with the Design Unit by means of the following tools: The Creative Brief, Project Plan, Requirements Analysis, Use Cases, Site Design, Site Architecture, Implementation Methodologies and Testing.

The Business Unit Interfaces with the Technical Unit by means of the following tools: Use Cases, Data Model, Business Processes and Testing.

b. Design

The Design Units are responsible for the creative development that accompanies project deliverables. This Unit consists of Web Designers, Senior Web Developers, Creative Directors, Interactive Producers, Web Developers and Graphic Designers.

The Design Unit Interfaces with the Business Unit using the tools described above. With the Technical Unit it uses the following tools: Project Plan, Technical Diagrams, Page

Architecture, Database Design, Reusable Apps, New Apps, Source Code, Equipment and Testing.

c. Technical

The Technical Unit is responsible for providing the guidance needed to select and acquire the technical necessities of the project and the Client deliverable. This unit consists of the following professionals: Technical Architects, Engineers, Database Admin, Programmers, Technical Writers and Documentation Managers.

As is illustrated by Venn Diagram (Figure 4), each professional uses his/her respective tools to produce project deliverables that lead to the final deliverables for the Client. These final deliverables include E-Commerce Functionality, Work Products, Project Workbook, Users Manual, Ops Manual, Training, Maintenance and Support and ROI.

4. Web Publishing Discipline – Applying Design to Website Development

Roles and Responsibilities – Key Roles

Senior Program Manager

The Principal (Sales Account Manager) - The Principal, or in some cases Sales Account Manager, is the individual who has overall responsibility for maintaining the professional relationship between your company and the Client. This person is typically responsible for monitoring and participating in multiple projects within one Client organization.

Program Manager

The Program Manager is a person responsible for managing the delivery of multiple projects and services with large project teams and full budgetary responsibility. This is an optional position, most appropriate when providing services to large-scale organizations where it is likely that multiple solution providers (i.e. subcontractors) are required in order to meet the Client's solution requirements and multiple projects are running simultaneously. A Principal may sometimes fulfill this role.

Interactive Producer

The Interactive Producer is a team leader/jr. project manager who has a Web design/creative background and generally comes from the design or business career path. Sometimes called an Information Architect. The producer determines the website content and works with the consultant to develop the branding issues for the site, coordinating resources concerning creative design and web front end (GUI) development. The interactive producer is more concerned with the graphic design and information architecture.

Consultant

The consultant functions as the business analyst for the Client. He/she ensures that the website reflects demographic, marketing and business case goals and constraints. He/she assists in testing and copywriting. He/she also writes the specifications for the site, works with the Client to define use cases and business rules. He/she is involved heavily with the requirements gathering phase and ensures that requirements are converted into effective use cases, collects requirements and hopes to move toward managing the customer relationship as an engagement manager or senior consultant.

Project Manager

The project manager schedules and controls all aspects of project work; tracks and reports on staffing, task status and all issues. The project manager can come from any career path and assists as necessary from his/her own area of expertise (technical, design or business). The project manager manages resources for the entire project including coordinating with the TPMs or Producers. He/she also provides in depth knowledge of the business problem, manages the project and supervises consultants

Developer

The developer usually has some technical skill set like Java, C++, JavaScript, etc. and implements the technical portion of the project, using the spec provided by the consultants. He/she codes the functional interactions and meets the programmatic needs of the website.

Technical Architect

The technical architect designs and implements the system architecture for complex websites. "System architecture" shows all the pieces and their interactions, including hardware, network, database, Web Sites and foundation software. He/she is assigned to a project to ensure adherence to the internal company infrastructure. He/she is a "developer" of internal infrastructure, helping to flesh out system architecture, and communication between the layers of the Web Site (database, ui, messaging, middleware, etc). He/she offers input into what's possible and what's not. He/she is also involved with due diligence selection of hardware and software for the project and in defining the Client solution from a technical perspective.

Programmer

The programmer writes the source code that is required by the various components of the site in order to ensure that they interoperate.

Technical Writer

The technical writer is responsible for producing the project workbook, user guide, glossary and related materials to be left with the Client at the close of the project. He/she works with the project manager to ensure that all work products are completed and properly documented. He/she edits all documents, proofreads and corrects to ensure that the deliverables are properly formatted for presentation to the Client at project close. He/she also interfaces with the various team members to ensure that all documentation is clear and properly formatted.

Documentation Manager

The documentation manager is responsible for the availability of knowledge to the team. He/she manages the project website and interfaces with the technical writer to ensure the proper collection, storage and availability of all documents needed for and produced by the project team. He/she monitors source code storage and versioning and ensures that the project workbook contains finished source code as well as historical work products that were developed throughout the project. In some cases, the documentation manager also assumes the role of the technical writer.

5. Work Products

Work Product 1.
New Client Acquisition. Client accepts proposal and requests creative work from Client. A senior level meeting from each of the three business units is held to identify resources and establish the development team. Senior level management identifies the senior project leads from each of the business units (Business, Design and Technical) and arranges a team-building meeting.

🟡 **Senior Business, Design, Technical Staff**

🔴 **ERA Solutions Framework**

🔵 **Team Leads**

Work Product 2.
Team Building Meeting. Senior level management and the senior project leads meet to identify resources available and select team members from each of the business units to

function as project creative people. Discuss major issues regarding the project, deliverables and business cases identified with the Client. Establish budget codes and budgetary guidelines and preliminary timeline.

● Senior Business, Design, Technical Staff, Team Leads

● ERA Solutions Framework

● Team Members, Budget & Finance Codes, Preliminary Timeline

Work Product 3.

Team Meeting. Establish project website and collaboration tools. Integrate change order management procedures and quality control and assurance procedures. Cover preliminary project plan with team members and ensure that the plan is workable. Develop Requirements Gathering work products and assign responsibilities and due date.

● Senior Business, Design, Technical Staff, Team Leads, Team Members

● ERA Framework, Budget and Finance Codes, Preliminary Timeline

● Project Website, Procedures, Requirements Gathering Work Products and Responsibilities

Work Product 4. Requirements Gathering

Client meeting(s) to discuss details. Producers, designers, writers and account manager meet with Client to get details about the project. This is the best time to gather information for each team involved. Even for simple prototypes, it is important to ask many questions of the Client including: what types of sites does the Client like; what tone do they prefer for their site; what colors do they like (bright, muted);

do they want a site that it is elegant, utilitarian, loud, bold; what is the main and most important function of the site such as to distribute information, to shop, etc.

Does the content exist but need re-purposing, or do they want you to create content for them? If you create content for them, is there a site they'd like to emulate in terms of style, tone, subject matter, organization, etc? We should create a standard list of questions for this meeting. This meeting could also serve as a pre-creative brainstorm with the Client.

The resulting work product is the Requirements document portion of the project workbook.

🟡 Senior Business, Design, Technical, Team Leads, Team

🔴 Use Cases, Budget and Finance Codes, Preliminary Timeline

🔵 Requirements Document

Work Product 5.

Document A: Project Brief. This document is created by the interactive producer (or the project manager) to give a brief introduction on the proposed project. It should include:

1. Client participants and their roles and responsibilities
2. Client contact information and roles
3. Brief description of the project
4. Project number; project billing code (if possible)
5. Rough schedule including a drop dead date
6. Initial resources needed

7. The purpose of the project (re: branding, new site, increase sales, etc.)
8. Location of all relevant information (on whose drive, where to find, etc.) so anyone can have access
9. Final deliverables determined
10. Technologies that will be used

Interactive Producer or Project Manager

Requirements Document

Project Brief

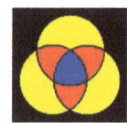

Work Product 6.

Internal project meeting. Corporate staff (interactive producers, business development account managers, writers, designers) that will be working on the project meet to discuss the scope of the project. The Project Brief is distributed and discussed.

Entire Team

Requirements Document, Project Brief

Discuss Scope of Project - Work Products

Work Product 7.

Document B: Project Plan. Interactive producers develop the project plan, basically a finalized version of the project brief. It includes:
1. Client participants and their roles
2. Client contact information and roles
3. Brief description of the project
4. Project number; project billing code
5. Final schedule including a drop dead date
6. Resources required

7. The purpose of the project
8. Location of all relevant information (on H drive) so anyone can access
9. Client should sign off on this document
10. Final deliverables determined
11. Technologies that will be used
12. Explain how a change order works and let Client know they how will be used. These will be generated throughout the project. Change order includes:
   -Description of change
   -Why we need to make the change
   -How the budget is affected
   -How the timeline is affected
   -Client signoff built in

● Interactive Producer

● Requirements Document, Project Brief, Work Products

● Project Plan

Work Product 8.

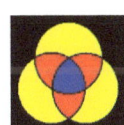 **Creative Brainstorm.** Interactive producers, writers, designers, and developers have a creative brainstorm about the project. They discuss various issues including: the theme of the site, what should be unique about the site, what functionality should be included in the site, what tone the site should convey including look and feel. Storyboard ideas could be elaborated in this step.

● Team Members

● Requirements Document, Project Brief, Work Products

● Refined Project Plan and Work Products

Work Product 9.

Document C: Creative Brief. This is an internal document to be created by the creative director and interactive manager. It should include:

1. Project background (an introductory paragraph about the project)
2. Scope, including:
   - Creative goal
   - Theme if there is one
   - Amount of content we have to generate vs. content provided
   - Artwork needed (illustrations, photos, etc.)
   - Reference materials (printed, URLs, etc.)
   - Branding
   - Style/look and feel
   - Voice
3. Deliverables that writers and designers are responsible for creating
4. Location and filing structure of creative files

🟡 Creative Director or Interactive Manager

🔴 Requirements Document, Project Brief, Work Products

🔵 Creative Brief

Work Product 10.

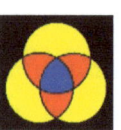

Document D: Site Map. Interactive producers create a site map for the Website to be developed. Client must sign off.

- Interactive Producer
- Requirements Document, Project Brief, Work Products, Creative Brief
- Site Map

Work Product 11.
Document E: Page Architecture. Designer and/or interactive producers create a document that outlines the page architecture.

- Designer or Interactive Producer
- Requirements Document, Project Brief, Work Products, Creative Brief Site Map
- Page Architecture

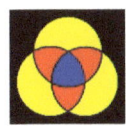

Work Product 12.
UI Brainstorm. Developers, designers and interactive producer meet to discuss UI solutions. This meeting specifies the functionality of the site. Questions such as "Will we use pull down menus, spawning browsers, rollovers, etc.?" will be explored. Also, navigation questions will be answered.

- Developers, Designer and Interactive Producer
- Requirements Document, Project Brief, Work Products, Creative Brief Site Map, Page Architecture
  User Interface and Navigation

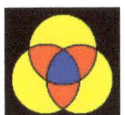

Work Product 13.
The Development Phase Begins. Everyone gets to work on making the plans a reality. Everyone receives copy of project schedule and timelines.

*

- **Team Members**
- Requirements Document, Project Brief, Work Products, Creative Brief
  Site Map, Page Architecture, User Interface and Navigation
  Source Code, Site Design

Work Product 14.
Status Meetings. Progress or status meetings between interactive producers, designers and writers to discuss how things are going and whether or not the creative team needs more information. Issues should be brought up and resolved. A meeting schedule can be established for the individual project.

- **Team Members**
- Requirements Document, Project Brief, Work Products, Creative Brief
  Site Map, Page Architecture, User Interface and Navigation,
  Source Code, Site Design
  Meeting Schedule, Issues and Change Order Decisions

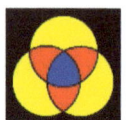

Work Product 15.
Client sees round 1 of designs- usually three designs with or without text in place. Client gives feedback to Team Leads, designer and producers. Client sign off.

- Client, Senior Staff, Team Leads
- Requirements Document, Project Brief, Work Products, Creative Brief
  Site Map, Page Architecture, User Interface and Navigation,
  Source Code, Site Design
  Client Feedback and Signoff

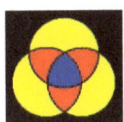

Work Product 16.

Second round of designs. Designers use feedback to create next round of designs. Client signoff.

Client, Senior Staff, Team Leads

Requirements Document, Project Brief, Work Products, Creative Brief
Site Map, Page Architecture, User Interface and Navigation,
Source Code, Site Design
Client Feedback and Signoff

Work Product 17.

Client sees first round of writing for the site (if applicable). Client gives feedback to creative director, writer and producers. Client sign off.

Client, Senior Staff, Team Leads, Writer/s

Requirements Document, Project Brief, Work Products, Creative Brief
Site Map, Page Architecture, User Interface and Navigation,
Source Code, Site Design, Recommended Site Content
Client Feedback and Signoff

Work Product 18.

Second round of writing. Writers get feedback and continue to refine the text. Client signoff.

Client, Senior Staff, Team Leads, Writer/s

Requirements Document, Project Brief, Work Products, Creative Brief
Site Map, Page Architecture, User Interface and Navigation,
Source Code, Site Design, Recommended Site Content
Client Feedback and Signoff

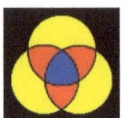

Work Product 19.

Site structure is built. Basic framework and templates are built according to site map. This should be tested by developers then reviewed by designer and producer, comparing it to site map. Once this is approved content can be pulled in.

● Developers, Designers, Creative Producer, Team leads

● Requirements Document, Project Brief, Work Products, Creative Brief
Site Map, Page Architecture, User Interface and Navigation,
Source Code, Site Design, Recommended Site Content
● Site Structure

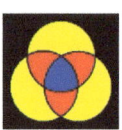

Work Product 20.

Final graphics and content delivered to developer(s). Everything starts coming together as the site look and feel takes shape. Explain SEO content strategies and META tags.

● Designers, Creative Producer, Team leads

● Graphic Design Work Products

● Final Graphics

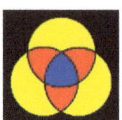

Work Product 21.

Graphics and content integrated into structure for working site. Preliminary site testing begins.

● Developers, Team leads

● Graphic Design Work Products

● Integration of Graphics into Site

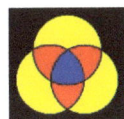

Work Product 22.

Testing. Done by the producers, designers and developers, testing the following:

1. Basic functionality (all links working, images loading)
2. Browser compatibility
3. Titles accurate on every page
4. Specified images in correct places

○ Team Members

● Pre-delivery E-commerce Site

● Fully Functional Site

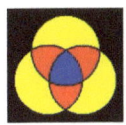

Work Product 23.
Final deliverables given to Client.

- All source and documents will be stored and versioned in a central filing area both online and hard copy
- A Project Workbook will be generated for each project. See Project Workbook Section of this document
- You will need to establish a nomenclature for electronic files and an owner should be assigned
- You should educate those who "sell" the various work products
- Client looks over site and gives authorization to go live or sets "go-live" date.

● Senior Staff, Team Leads

● E-commerce Site

● E-Commerce Functionality, Project Workbook, Users Guide, Ops Manual, Training, Maintenance and Support, ROI

6. Defining WDM Work Products - Object-Oriented Methodology

We have developed and refined an implementation approach that ensures that Client E-commerce sites will be implemented as designed -- and within a short timeframe. During the Project Initiation phase, which focuses on Requirements clarification, our approach employs a method of iterative prototyping where web system designers work with technical architects and business consultants to create an accurate user interface and robust back-end design blueprint for full support of the Client's vision.

Our Web Site Development methodology is designed to maximize the synergy among creative design, business strategy and technical architecture while retaining a strong deliverables focus. This methodology has demonstrated its usefulness in mitigating risk and managing the delivery of the final product. It combines years of experience in publishing Web sites and building large systems into a comprehensive model for a modern E-business solution.

Our carefully planned implementation approach will ensure that work is done with a minimum of cost, on time and under budget. Speed-to-market is improved which means the Client will be offering new functionality to site visitors sooner---and that means more revenue.

Finally, such issues as change order management and quality control and assurance are tightly managed, giving us the ability to minimize costly mistakes, reworking and retesting.

Project Management

This document recommends an "Object-Oriented" methodology. The object-oriented methodology is a widely used experience-based approach oriented around work products and centered on the production and development of the project workbook.

Instead of thinking in terms of a process, OOSD thinks in terms of real-world objects that act autonomously, yet in collaboration with other objects, to perform higher-level behavior. OOSD focuses on the artifacts or work products produced during a Web Site development project and the manner in which they are logically organized into the project workbook. It focuses on objects and their behavior.

Our methodology is built around a project website whose index tracks and links to specific work product deliverables. Collaboration Web Site enables timely communication with team members and Client management and facilitates quick decision-making and Client review. The document management system provides for collaboration, peering reviews and version tracking, providing at once, a status report, work schedule and document source. Project team members, Client as well as Client management can obtain a daily update of project status and issues by accessing one document structured around both the project schedule and workbook.

The project website should be structured so that each deliverable has a specific folder under its appropriate heading. This will enable the production of a cohesive product and help us identify needed deliverables and next steps as the documents unfold along with the project schedule.

Project Workbook

Every team member should understand the project workbook outline and what will be required as each performs assigned work products. Object-Oriented Web Site Development Methodology defines and clarifies the documentation requirements for each work product. This enables us to develop a consistent approach wherein each work product has its own specific set of requirements, procedures, templates, worksheets and samples. This will also create greater project cohesion among the various team members — particularly if some of the assigned individuals change in mid-project. Such a workbook will not only communicate the high standards required of each project but enable a faster speed-to-market for each successive project - as each team member becomes accustomed to providing deliverables in a uniform way across a whole range of projects - giving Client a strong discriminator in the marketplace.

Project Workbook Outline

Task Categories	Deliverables
1.0 Set Scope of Project — Project Plan	1.0 Project Plan 1.0.1 Client participants and their roles and responsibilities 1.0.2 Client contact information and roles 1.0.3 Brief description of the project 1.0.4 Project number; project billing code (if possible) 1.0.5 Rough schedule including a drop dead date 1.0.6 Initial resources needed 1.0.7 The purpose of the project (re: branding, new site, increase sales, etc.) 1.0.8 Location of all relevant information so anyone can access 1.0.9 Final deliverables determined 1.0.10 Technologies that will be used 1.0.11 Change Order Management

| 2.0 Requirements Analysis
○ 2.0.1 Requirements Gathering
○ 2.0.2 Project Management
○ 2.0.3 Problem Analysis
○ 2.0.4 User Interface Model
○ 2.0.5 Screen Flows
○ 2.0.6 Screen Layouts
○ 2.0.7 User Interface Prototype | 2.0.1 Requirements Gathering – Functional/Non-Functional
• 2.0.1.1 Strategy
• 2.0.1.2 Problem Statement
• 2.0.1.3 Use Case Model/Diagram
• 2.0.1.4 Nonfunctional Requirements
• 2.0.1.5 Prioritized Requirements
• 2.0.1.6 Business Case
• 2.0.1.7 Acceptance Plan
2.0.2 Project Management
• 2.0.2.1 Project Workbook Outline
• 2.0.2.2 Intended Development Process
• 2.0.2.3 Resource Plan
• 2.0.2.4 Schedule
• 2.0.2.5 Release Plan
• 2.0.2.6 Quality Assurance Plan
• 2.0.2.7 Risk Management Plan
• 2.0.2.8 Reuse Plan
• 2.0.2.9 Test Plan
• 2.0.2.10 Metrics
• 2.0.2.11 Dependencies
• 2.0.2.12 Issues
2.0.3 Problem Analysis
• 2.0.3.1 Guidelines
• 2.0.3.2 Subject Areas
• 2.0.3.3 Object Model
• 2.0.3.4 Scenarios
• 2.0.3.5 Object Interaction Diagrams
• 2.0.3.6 State Models
• 2.0.3.7 Class Descriptions
2.0.4 User Interface Model
• 2.0.4.1 Guidelines
• 2.0.4.2 Screen Flows
• 2.0.4.3 Screen Layouts
• 2.0.4.4 User Interface Prototype |
| 3.0 Design
○ 3.0.1 Design Guidelines | 3.0.1 Design Guidelines
3.0.2 System Architecture |

	o 3.0.8.3 Assumption
	o 3.0.8.4 Outcome
	o 3.0.8.5 Description
	3.0.9 Design State Models
	3.0.10 Design Class Descriptions
	o 3.0.10.1 Description
	o 3.0.10.2 States
	o 3.0.10.3 Relationships
	o 3.0.10.4 Public Members
	o 3.0.10.5 Protected Members
	o 3.0.10.6 Private Members
	o 3.0.10.7 Notes
	3.0.11 Rejected Design Alternatives
	o 3.0.11.1 Description
	o 3.0.11.2 Context
	o 3.0.11.3 Assumptions
	o 3.0.11.4 Alternatives
	o 3.0.11.5 Considerations
	o 3.0.11.6 Decision
4.0 Resources	4.0 Resource Assignment Worksheet
5.0 Work Product Planning • 5.0.1 Set Project Increments • 5.0.2 Define Work Products • 5.0.3 Work Product Descriptions • 5.0.4 Schedule Work Products	5.0.1 Project Increments Worksheet • 5.0.1.1 Purpose • 5.0.1.2 Scope • 5.0.1.3 Guidelines • 5.0.1.4 Process 5.0.2 Work Product Worksheet • 5.0.2.1 Identifier • 5.0.2 2 Date • 5.0.2.3 Author • 5.0.2.4 Owner • 5.0.2.5 Status • 5.0.2.6 Issues • 5.0.2.7 Metrics • 5.0.2.8 Traceability • 5.0.2.9 History 5.0.3 Work Product Descriptions 5.0.4 MS Project Work Schedule • Activities • Start Dates and Durations for Each Activity • Work Assignments • Milestones

	• Resources • Costing

6.0 Development Phase
  ○ 6.0.1 User Interface Design
  ○ 6.0.2 System Design
	• 6.0.1 User Interface Design
	• 6.0.2 System Design
	  ○ 6.0.2.1 Platforms
	  ○ 6.0.2.2 Languages
	  ○ 6.0.2.3 Performance
	  ○ 6.0.2.4 Interoperability
	  ○ 6.0.2.5 Persistence
	  ○ 6.0.2.6 Maintainability
	  ○ 6.0.2.7 Use of standard components and subsystems
	  ○ 6.0.2.8 Reuse
	  ○ 6.0.2.9 Cost
	  ○ 6.0.2.10 Time

7.0 Implementation
  ○ 7.0.1 Coding Guidelines
  ○ 7.0.2 Physical Packaging Plan
  ○ 7.0.3 Development Environment
  ○ 7.0.4 Source Code
  ○ 7.0.5 User Support Materials
	7.0.1 Coding Guidelines
	  ○ 7.0.1.1 File Naming Conventions
	  ○ 7.0.1.2 File Structure
	  ○ 7.0.1.3 File and Function prologues (Security and Copyright)
	  ○ 7.0.1.4 Identifier Naming Conventions
	  ○ 7.0.1.5 Global Names
	  ○ 7.0.1.6 Class/Structure Layout
	  ○ 7.0.1.7 Initialization
	  ○ 7.0.1.8 Use of Types within Language
	  ○ 7.0.1.9 Calling Conventions and Return Types
	  ○ 7.0.1.10 Cohesion, Encapsulation, Binding
	  ○ 7.0.1.11 Memory Management
	  ○ 7.0.1.12 Exception and Error Handling
	  ○ 7.0.1.13 Use of Language-specific Features
	  ○ 7.0.1.14 Terseness of Expression
	  ○ 7.0.1.15 Performance
	  ○ 7.0.1.16 Portability
	  ○ 7.0.1.17 Formatting

Other Books by Robert Villegas

Poetic Prose and Poetry
The Raven Haired Girl
The Boy Who Stood Alone
Crushing the Alinsky Radicals
Unkilling Jesus
Adam Reborn and Adam Rayberne
Hospitality Event Planning Handbook
How to Write a Sponsorship Proposal
The Sport Sponsor Handbook
Finding Sponsorship

To order:
http://www.documentservicesinternational.com/books.html

www.ingramcontent.com/pod-product-compliance
Lightning Source LLC
Chambersburg PA
CBHW040819200526

45159CB00024B/3048